W9-BEK-430

An Optical Artist

Exploring Patterns and Symmetry

Greg Roza

PowerMath™

The Rosen Publishing Group's

New York

Published in 2005 by The Rosen Publishing Group, Inc.
29 East 21st Street, New York, NY 10010

Book Design: Daniel Hosek

Photo Credits: Cover M.C. Escher's *Hand with Reflecting Sphere* © The M.C. Escher Company—Baarn—Holland. All rights reserved.; p. 5 © William Whitehurst/Corbis; p. 6 (top) M.C. Escher's *Drawing Hands* © The M.C. Escher Company—Baarn—Holland. All rights reserved.; p. 6 (bottom) M.C. Escher's *Self Portrait* © The M.C. Escher Company—Baarn—Holland. All rights reserved.; p. 9 © Joseph Sohm/ChromoSohm Inc./Corbis; p. 13 © Ralph A. Clevenger/Corbis; p. 15 M.C. Escher's *Symmetry Drawing E42* © The M.C. Escher Company—Baarn—Holland. All rights reserved.; p. 16 © Francesco Venturi/Corbis; p. 17 © Roger Wood/Corbis; p. 18 © Hubert Stadler/Corbis; p. 19 © Phillipa Lewis/Edifice/Corbis; p. 21 M.C. Escher's *Liberation* © The M.C. Escher Company—Baarn—Holland. All rights reserved.; p. 22–23 M.C. Escher's *Day and Night* © The M.C. Escher Company—Baarn—Holland. All rights reserved.; p. 24 detail from M.C. Escher's *Day and Night* © The M.C. Escher Company—Baarn—Holland. All rights reserved.; p. 27 M.C. Escher's *Reptiles* © The M.C. Escher Company—Baarn—Holland. All rights reserved.; p.28 M.C. Escher's *Sun and Moon* © The M.C. Escher Company—Baarn—Holland. All rights reserved.; p. 30 M.C. Escher's *Two Intersecting Planes* © The M.C. Escher Company—Baarn—Holland. All rights reserved.

Library of Congress Cataloging-in-Publication Data

Roza, Greg.
An optical artist : exploring patterns and symmetry / Greg Roza.
p. cm. — (PowerMath)
Includes index.
ISBN 1-4042-2927-2 (library binding)
ISBN 1-4042-5117-0 (pbk.)
6-pack ISBN 1-4042-5118-9
1. Escher, M. C. (Maurits Cornelis), 1898-1972—Juvenile literature. 2. Tessellations (Mathematics)—Juvenile literature. 3. Mathematics in art—Juvenile literature. I. Title. II. Series.

NE670.E75R69 2005
769.92—dc22

2004000015

Manufactured in the United States of America

Contents

M.C. Escher—An Optical Artist

Have you ever heard of Optical Art? The word "optical," which means relating to the eye, may give you the clue that this kind of art focuses more on how you see the work than on what the subject of the work is. Optical Art is a form of art in which colors and shapes are used to create visual effects. For instance, look at the picture on the opposite page. Depending on how you look at the chessboard, you may seem to be looking on the inside or outside of a cube-like structure. It's almost like the picture plays tricks on your eyes. Optical artists try to create similar confusing, exciting, or exaggerated effects in their works. How do they do this? You may be surprised to learn that Optical Art relies greatly on mathematics.

To many people, mathematics and art probably seem very different. They may think of mathematics as factual and bound by rules, and art as personal and creative. However, some mathematicians have shown the world that math can be imaginative. Some artists have also proven that art can be mathematical. In fact, without math we would not have many works of art.

Symmetry, for example, plays an important role in many works of art. Symmetry is a branch of geometry that deals with shapes and objects that look the same on both sides of a centerline. For instance, if you drew a line through the middle of 1 of the chess pieces from top to bottom, each side would be a mirror image of the other. Some artists are well known for their use of symmetry in their work. Perhaps no artist is better known for this than the Dutch artist M.C. Escher.

Optical Art began to be called "Op Art" in the 1960s as many artists began to experiment with different visual effects of the kind that M.C. Escher had been applying to his works for years.

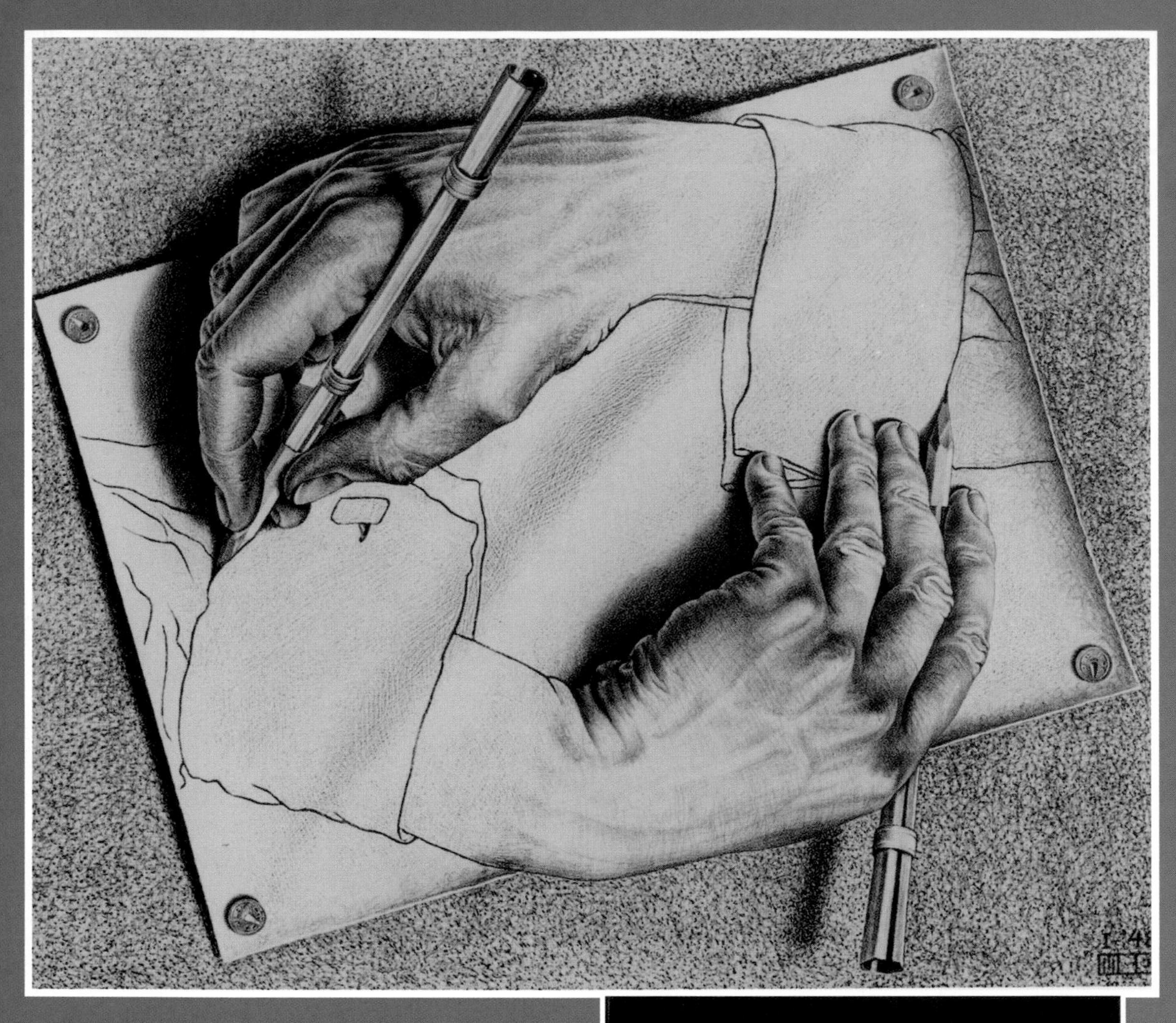

In the 1943 self-portrait to the right, Escher appears to be spying on his audience rather than posing for them as in a traditional self-portrait. In 1948, he created *Drawing Hands,* shown above, which shows his unusual style of combining realistic images with fantasy.

Maurits Cornelis Escher was born in Leeuwarden, Netherlands, in 1898. The Kingdom of the Netherlands, which is sometimes called Holland, is a country in western Europe. At a young age, Escher was encouraged by his father to become an **architect**. In 1919, Escher attended a school for architecture and art in Haarlem, Netherlands. When one of Escher's teachers saw his drawings, he convinced Escher to become a student of drawing and printmaking.

Much of Escher's early work included views of nature, like trees, flowers, and animals. Escher also developed an interest in drawing reflections, symmetrical shapes, and patterns. In 1922 and 1936, Escher visited a fourteenth-century **Muslim** palace in southern Spain called the Alhambra (al-HAM-bruh). Escher sketched hundreds of patterns he found in this palace.

Escher was in awe of the geometric elements of the Alhambra but wanted to bring them to life somehow. Upon seeing his sketches, Escher's brother gave him some papers to read about symmetry and mathematics. Escher drew on these ideas and the **Islamic** patterns of the Alhambra sketches as he began to approach his work with a mathematical eye. The result was a new kind of art. Some of Escher's works were even shown in science museums as well as art museums. Mathematicians and art lovers alike soon took notice of his original style, which often involved the creation of symmetrical patterns known as tessellations.

What Is a Tessellation?

A tessellation is an arrangement of similar shapes on a plane. The shapes fit together perfectly without overlapping or leaving gaps. This arrangement of shapes forms a pattern that can be extended **infinitely** in every direction on a plane. One of the simplest examples of a tessellation is a checkerboard. The squares fit together perfectly with no gaps and without overlapping.

Tessellations are sometimes called tilings, which is a fitting name since they can look like floor or wall tiles. The tile surfaces in some kitchens, bathrooms, and public buildings are good examples of tessellations in the world around us. Some tile surfaces use the repetition of a single shape, like the squares of the floor in the picture on page 9. Other tile surfaces may be more complex and involve the repetition of 2 or more shapes. These tiles fit together perfectly, do not overlap or leave gaps, and can be extended as far as they are needed.

Other tessellations are made up of numerous irregularly shaped pieces that fit together perfectly to tile a plane. A jigsaw puzzle, for example, can be made up of dozens, hundreds, even thousands of pieces, each different from the rest. Every piece must be placed perfectly for the puzzle to work. When each piece is in its proper place, a whole picture is created with no gaps or overlapping between pieces.

Shapes are sometimes said to "tile the plane" if they can be extended infinitely on a plane with no gaps or overlapping.

Tessellations are often examples of **symmetry transformation**. Symmetry transformation is the process of copying a shape and moving it to another location. Transformations include shapes that have been repeated and flipped, turned, or slid.

In this tessellation, many squares of the same size are tiled across the page. The original shape, a single square, was repeated and slid to a new location. This process was repeated many times.

In many tessellated patterns, shapes come together to form a perfect fit at a **vertex**. The angles around any vertex add up to 360 degrees (or 360°), which is the number of degrees in a circle. A square, for example, has four 90-degree angles. When four squares are fitted together, the angles around the vertex add up to 360°.

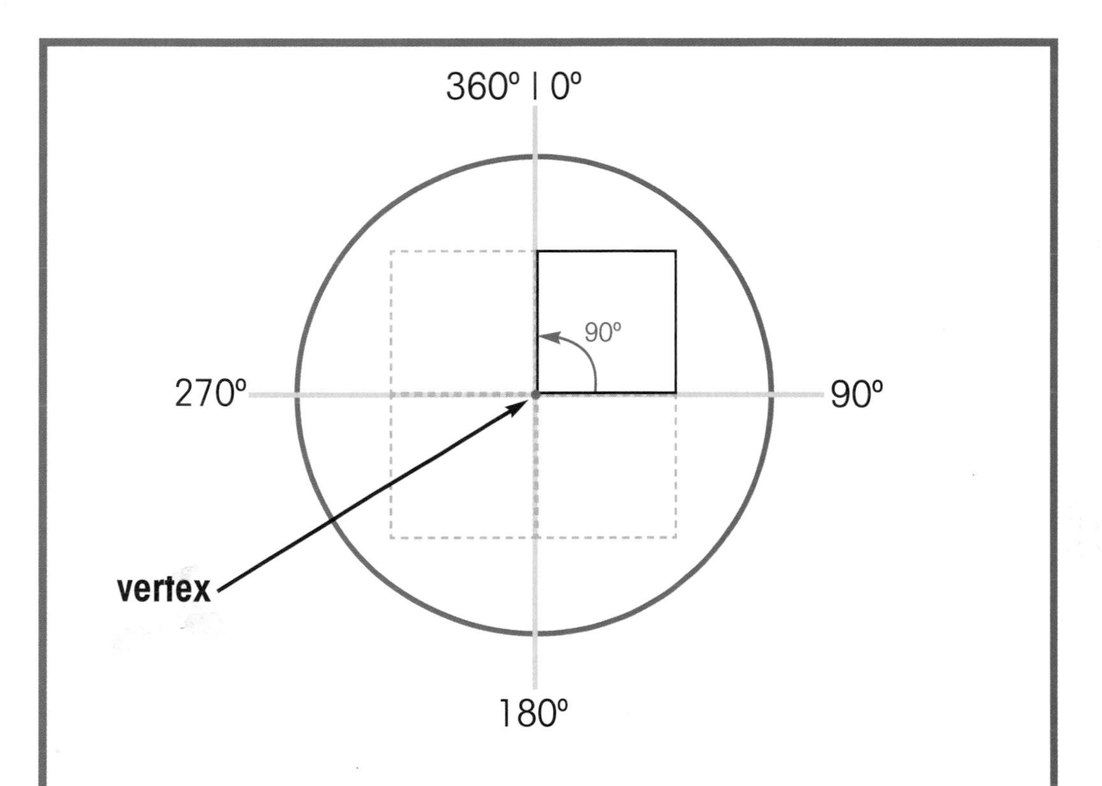

Tessellations are patterns of repeated shapes. Notice in this example that the original square shape was repeated several times and turned around the vertex until it formed 4 identical squares.

The most basic tessellations are called regular tessellations. Regular tessellations are made up of regular **polygons**. A regular polygon is a shape whose sides are all the same length and whose angles are all the same size. As we have already seen, a square can be used to form a regular tessellation. Four squares can be fitted together around a vertex without overlapping and without leaving any gaps. A square can be tiled infinitely across a plane.

An equilateral triangle has 3 angles that are 60°. The picture below shows that a total of 6 equilateral triangles can be fitted together around a vertex with no overlapping and no gaps. The 6 angles all add up to a total of 360°.

Only one other regular polygon can be used to make a regular tessellation: a regular **hexagon**. A regular hexagon has 6 angles that are all 120°. Three regular hexagons fit snugly together around a vertex, and 120° multiplied by 3 is 360°. The shape formed by the 6 triangles to the right is a regular hexagon.

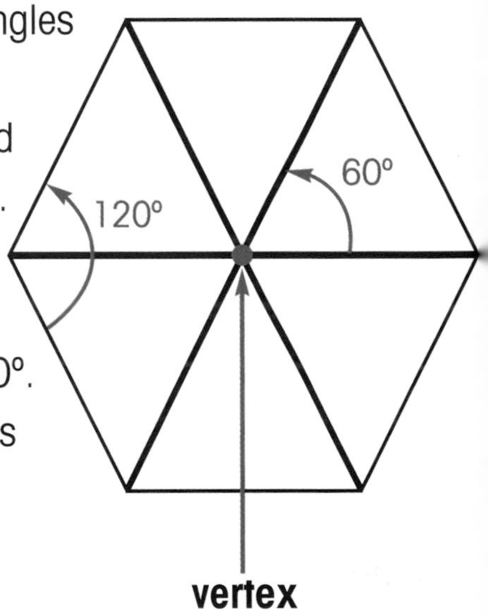

60°
120°
vertex

Honeybees build wax nests called honeycombs to hold their young and their supplies of honey. A honeycomb is an example of a regular tessellation that can be found in nature. Each cell in the honeycomb is a regular hexagon.

120°

The 3 kinds of regular tessellations are made up of equilateral triangles, squares, or regular hexagons. However, tessellations can be made with 3 or more regular polygons as well. These tessellations are called semiregular tessellations. There are 8 forms of semiregular tessellations.

The first example, shown below, is one of the simplest semiregular tessellations, which is made up of 2 regular **octagons** and a square. The sides of the octagons are the same length as the sides of the square. Notice that these shapes fit perfectly around the vertex with no gaps or overlapping. The angles of a regular octagon are all 135°. As we saw earlier, the angles of a square are all 90°. So, 135° added to 135° added to 90° equals 360°.

Irregular (or nonregular) tessellations are the tiling of complex shapes. Irregular tessellations can include a large number of shapes that may or may not be symmetrical, as long as they fit together with no gaps or overlapping.

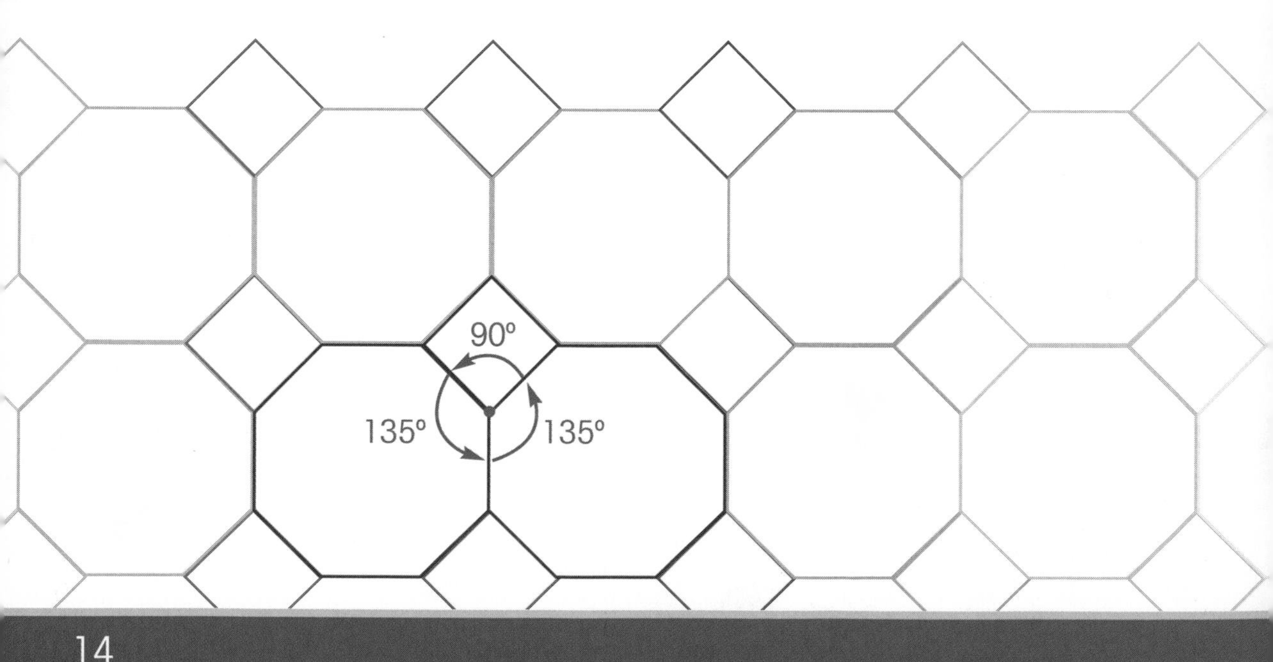

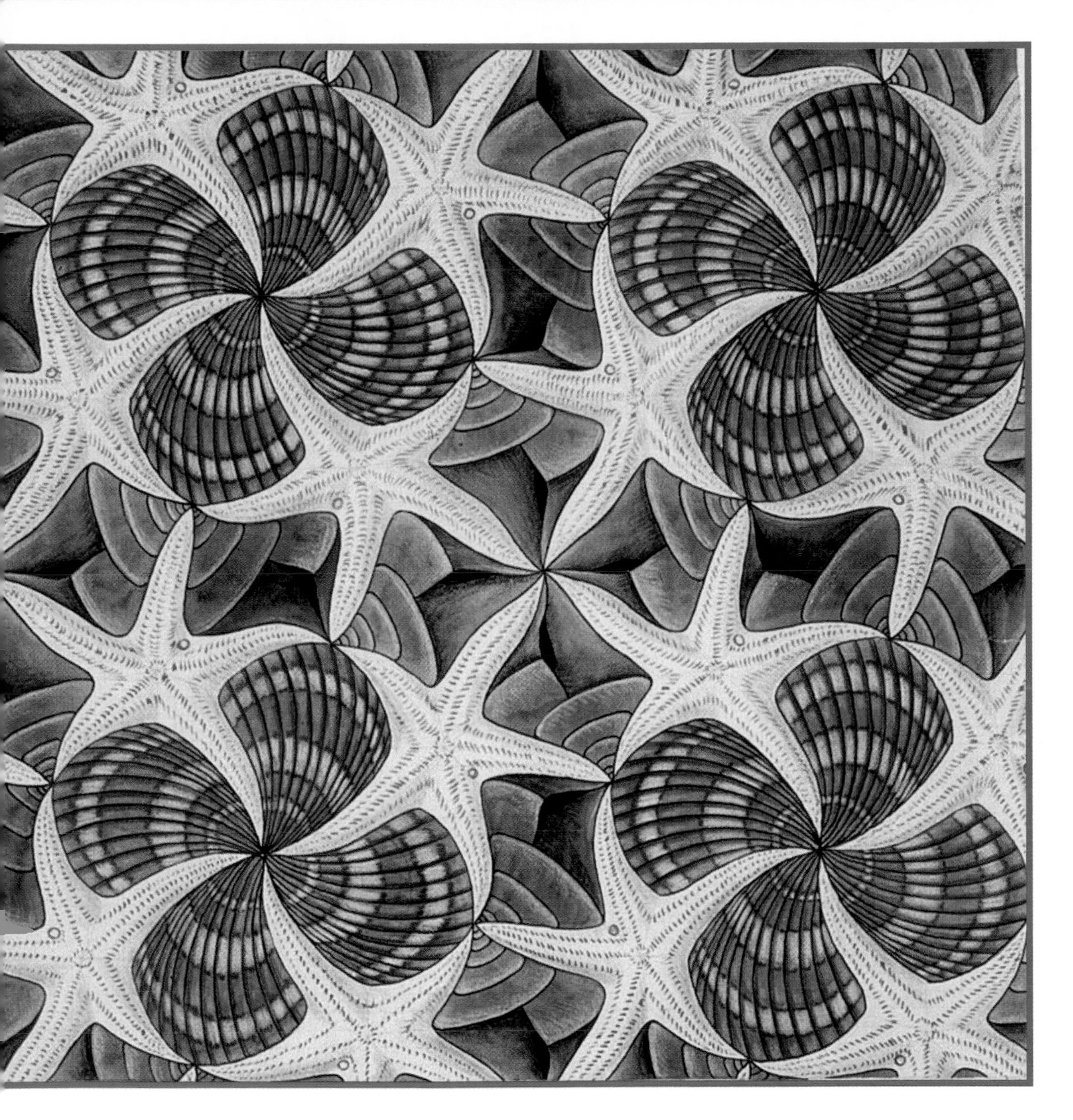

Escher became a master at creating complex irregular tessellations with shapes and figures, such as the shells and starfish above in *Symmetry Drawing E42*.

Islamic Art and the Alhambra

Tessellations have appeared in the art and buildings of ancient cultures around the world. We have found examples of tessellations and symmetry in Egyptian tombs, Roman mosaics, Asian rugs, African tapestries, and Native American clothing as well as in many other cultures.

Even though the Islamic tiles shown above have different decorations on them, they are all regular hexagons forming a regular tessellation.

The Muslim culture is widely known for its use of symmetry and tessellations in works of art. Islamic **mosques** and buildings are often covered with **ornate** tiles that form complex tessellations. The symmetrical patterns of Islamic art represent the complex and infinite nature of God, whom Muslims call "Allah." Muslims represent God this way in their art because they do not believe God should be shown as a living being. Muslims have also used symmetrical patterns to decorate clothing, rugs, ceramics, books, and metalwork.

This dome in Mahan, Iran, lies over the tomb of an important figure of Islam, Shah Nematollah Vali, who died in 1431.

Many believe that Muslims have built some of the most beautiful buildings in the world. Some consider the Alhambra palace to be the most beautiful example of Islamic art. The Alhambra, located in Granada, Spain, was originally built in the fourteenth century, but has undergone many changes over the years. This fortress and palace was used by several Muslim rulers and was large enough to be considered a walled city.

The Alhambra is filled with wood and stone carvings of geometric shapes. The tile walls, ceilings, and floors of the Alhambra feature complex symmetrical patterns. The Alhambra was neglected for many years and was even nearly destroyed. Today it is a monument to the artistic genius of the Muslim builders who created it. It is also a fascinating example of the use of geometric shapes and tessellations in art and architecture.

The name "Alhambra" comes from an Arabic word which means "red castle." Some people say that the castle may have gotten this name because the walls appear reddish as the sun sets. Others say that parts of the palace may have been built by the light of torches.

These walls found in the Alhambra were decorated in the fourteenth century. The designs are an example of the complex geometric patterns that can be found in the palace.

Escher, Art, and Math

Escher's visits to the Alhambra palace changed the way he looked at the world. It also changed the way he created art. Escher once said his trips to the Alhambra were "the richest source of inspiration I have ever tapped." Escher also thought it was a shame that Islamic builders and artists were forbidden by religious law to represent living creatures in their art. Escher respected this viewpoint, but he felt that something was missing from the complex patterns that he found at the Alhambra.

Escher began experimenting with tessellations. He modified the symmetrical patterns until they resembled something familiar and alive. Escher combined the natural world, with all of its constant movement and changes, with the world of mathematics, which to many had seemed unchanging and bound by rules. The results were new and remarkable, which raised the interest of many artists and mathematicians around the world.

The picture on the opposite page is titled *Liberation*. "Liberation" means the process of being set free. The triangles at the bottom of the work are gradually changed into birds and set free, or liberated. Perhaps Escher meant for this picture to be a symbol of the liberation of mathematical and artistic ideas from traditional forms.

Escher experimented with contrasting colors as well as figures. In *Liberation* on the opposite page, notice that the black and white birds in the middle take on softer, more natural colors as they come to life at the top.

Escher did not just create tessellations, he brought tessellations to life. Escher transformed basic geometrical shapes into lively animals and creatures. His creations looked like they flew right out of the plane on which he drew them. Remarkably, these creatures still had the symmetrical features of the tessellations from which they were created. Escher called these unusual transformations "**metamorphoses**" (meh-tuh-MORE-fuh-seez).

The picture below, titled *Day and Night,* is one of Escher's first experiments with tessellations. He created this work of art in 1938, after his second visit to the Alhambra. The entire picture is a good example of a symmetry transformation. While not exact, notice that the 2 halves of the print are nearly mirror images of each other, as if the image on the left was epeated and then flipped to create the image on the right.

"Transformation" and "metamorphosis" both mean a change of form.

At first, it is difficult to understand what is happening in the picture. This is because Escher uses common images to create a fantasy. The square plots of land at the bottom of the picture resemble a regular tessellation. However, as we move up the picture, the squares are transformed into birds!

The birds themselves on both halves of the picture undergo a transformation as we look from one side to the other. The light between the black birds on the left becomes the white birds on the right; the shadows between the white birds on the right become the black birds on the left. Both sets of birds are puzzle pieces that meet in the middle of the picture and fit together perfectly. No gaps and no overlapping are visible between the birds, just as shapes in a tessellation have no gaps or overlapping. *Day and Night* shows how Escher was able to bring his art to life.

The birds in *Day and Night* form an irregular tessellation. This is because even though they tile the plane perfectly, they are not regular polygons. However, it is interesting to note that the fields below the birds resemble a regular tessellation—a checkerboard pattern of squares. Just like *Liberation,* the regular polygons become the shapes of natural creatures. The squares at the bottom transform into the birds at the top of this work.

Escher believed that birds had a natural shape for forming tessellations. Can you think of any other animals that could form a tessellation?

Escher was not content to simply work with basic patterns. He wanted to breathe life into his patterns and allow them to leap right out of the plane on which they were created. He was able to do this because he had the ability to create incredibly lifelike drawings of **three-dimensional** objects. This means that the objects appear to have height, width, and depth, like objects we encounter in life.

In a work called *Reptiles,* shown on page 27, Escher used a tabletop with everyday objects as a setting for an unusual event. Near the center of the picture is a pad of paper. The drawing paper features a tessellated pattern of lizards. Each lizard is identical to the others, and they fit snugly together with no gaps and no overlapping. What makes the picture so unusual is that the lizards break free of the paper. The lizards crawl off the paper and across the table, before crawling back onto the page to join the flat **two-dimensional** lizards—who have only height and width—on the other side of the drawing!

The lizards on the drawing paper in *Reptiles* form a complex irregular tessellation. Just as in *Day and Night,* however, Escher used a regular tessellation—regular hexagons this time—as a pattern for the arrangement of the lizards. In fact, Escher left the hexagon tessellation in the drawing for viewers to see. Can you find it?

When describing *Reptiles,* Escher said, "Evidently one of the reptiles has tired of lying flat and rigid amongst his fellows, so he puts one plastic-looking leg over the edge and wrenches himself free." This quote and *Reptiles* itself are examples of Escher's unusual imagination at work.

Sun and Moon, pictured above, creates an illusion of depth. Each set of birds, dark and light, can be seen as a background for the other set.

Escher's pictures are often more complicated than they appear at first. *Sun and Moon,* shown on page 28, features a bird pattern. Like the other Escher patterns we have seen, this is an irregular tessellation. The bird shapes are not regular polygons, but they do tile the plane with no gaps and without overlapping. The difference between *Sun and Moon* and the other works we have seen is that each of the birds has its own special shape. This shows us that the shapes used in irregular tessellations do not need to be identical.

Although it is more difficult to see, Escher again used a regular tessellation as the basis for this picture. Notice that there are 4 rows of birds, and that the birds alternate between light colors and dark colors. This is similar to a checkerboard, which, as we saw earlier, is a regular tessellation using squares.

Sun and Moon is also more than a picture of birds. As the title suggests, it is also a picture of a sun and a moon. The light-colored birds show a sun and its rays reaching out toward the edges of the picture. The dark-colored birds show a moon surrounded by stars. Do you remember the definition of Optical Art from the beginning of this book? When creating *Sun and Moon,* Escher focused more on how you see the birds—as a sun or moon—rather than on the birds themselves.

During his lifetime, Escher's work redefined the way people thought about math and art. Mathematicians discovered new uses for mathematical concepts through Escher's artistic creations. Artists found creativity and beauty in Escher's mathematical patterns. Many became inspired to create Optical Art pieces themselves that were visually appealing and entertaining.

Escher once said, "Although I am absolutely innocent of training or knowledge in the exact sciences, I often seem to have more in common with mathematicians than with my fellow artists." Despite this view of himself, Escher is today remembered as a mathematical artist, as well as an artistic mathematician.

Glossary

architect (AHR-kuh-tekt) A person who designs buildings.

hexagon (HEK-suh-gahn) A shape with 6 angles and 6 sides.

nfinitely (IN-fuh-nuht-lee) Without an end.

slamic (ihz-LAH-mihk) Having to do with the religious faith of the Muslim people.

metamorphosis (meh-tuh-MORE-fuh-sihs) A change in form.

mosque (MAHSK) A Muslim place of worship.

Muslim (MUHZ-luhm) Having to do with the religion of Islam.

octagon (AHK-tuh-gahn) A shape with 8 angles and 8 sides.

ornate (ohr-NAYT) Elaborately decorated.

polygon (PAH-lee-gahn) A closed plane shape with straight sides.

symmetry (SIH-muh-tree) The state of having the same size and shape on opposite sides of a dividing line or around a point or axis.

symmetry transformation (SIH-muh-tree trans-furh-MAY-shun) Symmetry created by repeating a shape or object and then flipping, turning, or sliding the second shape to a new location.

three-dimensional (THREE–duh-MEN-shuh-nuhl) Having height, width, and depth.

two-dimensional (TOO–duh-MEN-shuh-nuhl) Having height and width.

vertex (VUR-teks) The common endpoint of the sides of an angle.

Index